THE ODOR OF LOVE

AND OTHER AROMAS

Text by Michael Dubson Sage
Art by Paul Theodore

Teddy Bear Books

TEDDY BEAR BOOKS
are published by
T & T Publications
Box 23
Boston, Massachusetts 02124

ISBN: 0-9658977-0-2

First Printing October, 1997
Printed in the United States of America
by Bookmasters, Inc., Mansfield, Ohio 44905

For Linda Werbner--who started it all--
But most of all, for Tim
MDS

For Dad, Dottie, the Duchess
and, of course, Joey
PT

CONTENTS

INVOCATION

ALTHOUGH I'm not quite another Shakespeare--
In fact, I'm not quite a Rod McKuen--
I've had quite my share of guilt, pain and fear
But I've not let it drive me to ruin

I'VE survived because I don't hold it in
And there is no real reason why I should
Not when there is my waiting ink and pen
To record my woes on dead, processed wood

IN swift meter and wicked metaphor
and with symbols that will outlast Stonehenge
I attack my pain as my words wage war
and the finished product my sweet revenge

I WRITE away my hurt; that'll show 'em
I put the pains of life into poem

THE REEK OF ROMANCE:

THE ODOR OF LOVE

LOOKING FOR LOVE

You're gone now
And I stay in bed
Thinking of your face and smile
As Geraldo carries on
In the next room
and the tea kettle
Boils on the stove

Our first night together
How I loved you
Now all that's left
Is the smell of your hair
On my pillow
And the stains of our evening
On my designer sheets

I can't help but wonder
If I'll ever see you again
Because I don't remember
If I wrote down your phone number
On our pizza-stained napkin
Or what your name was

WHEN I THINK OF YOU

When I think of you
I find myself singing
Olivia Newton-John
and Barry Manilow songs
And thinking of pretty flowers
Crashing waves and sandy shores
Hot dogs dripping mustard
Into the warm summer air

When I think of you
I find myself thinking of
Doris Days and Lancelot Knights
Heartfelt promises
And cologne commercials
Candy hearts and Valentine cards
Eurphoric, sun-drenched mornings

When I think of you
I find myself thinking of
World peace and chocolate kisses
Soft clouds and cotton candy
Teddy bear picnics
And Raggedy Anne parties
Wet dreams and hard bodies
Satin sheets and Trojan conquests
And happily ever afters

And I hope we can keep
Seeing each other
So I'll know if
I'll still love you this much
Once I really get to know you

WHEN I THINK OF YOU II

When I think of you
I find myself singing
Olivia Newton-John
And Barry Manilow songs

And then I stop thinking of you
and come home from work
And do the dishes
and take out the trash
and wish to hell
You'd help around the house

YOU LAZY
SCREWED UP
INCONSIDERATE
SACK OF SHIT!

I LOVE YOU SO MUCH

I love you so much
I don't mind it when

You floss your teeth in front of me
Wear your underwear for a week
You pick your nose in front of company
And don't listen when I speak

You clip your gnarly fingernails
At our sunny breakfast table
You fill our evenings with neurotic wails
Because you believe your persecution fables

You're always too tired
Then hot to trot when I'm asleep
You're always too wired
And you want me to be your pit bull, teddy bear and sheep
You flop around the house all day
You say you're depressed, not lazy
You always insist on getting your way
AND I SWEAR YOU'RE COMPLETELY CRAZY!!!

WHY CAN'T I FIND LOVE?

On it goes and still I'm alone
And I just don't know why I can't find love

You were so cute and sweet and an awful lot of fun
But you weren't ready for love, and before it started, it was over and done

You were able to love, and we happened to meet at a lonely junction
But hot sex isn't enough with a mess who just can't function

You say you want me, and you know I'm really hot for you
Your old lover holds you at home, so there's nothing we can ever do

You are sane, fun and free with arms for many a tender day
But when you held your love out to me, I smiled and turned away

And so it goes and still I'm alone
and I just don't know why I can't find love

I THINK I'M GONNA PUKE

Because you're so sweet and warm
You always reminded me of food
You're definitely delectable
I've so wanted to
Experience with you
What I've only known with
Melting ice cream and popsicles
Wiping my lips afterwards
With your satisfied smile

It was in the dark days of August
In the Dark Shadows of my favorite bar
You said you weren't interested
Oh yes, we both know that my table is already set
But lately it seems the casserole dishes are always empty
And I'm always dining alone
Now it feels like I've reached for the silver spoon
On an all-you-can-eat buffet
And missed it by one ugly inch

AND I THINK I'M GONNA PUKE!!!
Because I never got to feel your naked, freckled flesh
Pressed tight against mine
Under the full moon
After the Stevie Nicks concert

AND I THINK I'M GONNA PUKE!!!
Because I still don't know why you said no that evil evening
Was I too married? Did I want out too much?
Too strong, secure and successful for you?
Or just too short, ugly, old and fat?

But time goes on
And one day the phone rings
It's a friend of mine
Who calls often with his woes
Lamenting his miserable life
As he whines through his nose
He's gotten to feel your naked flesh
He's held you tight against the rain
Something magic has happened for him
His joy now doubles and triples my pain

AND I THINK I'M GONNA PUKE!!!
Because the two of you together makes me sick
Fool, you missed out on a really good deal
But you're never going to see me beg
You walked away from a full course meal
To gnaw on a neurotic chicken leg!!!

WHAT WOULD BARNABAS COLLINS HAVE DONE?

These days...
True loves are few and far between
So I cling to you with a deathlike grip
As I squirm with spasms of love obscene
You keep the hungry smile tight on my lips

I'm fervently obsessed with you
Not even centuries will end my need
You've got me in chains; I don't know what I'll do
I want to break free, grab you and make you bleed

Lonely, I hide in darkness and shame
And rail against the one I need so much
Cursed and alone, I search for blame
And I wait, not yet alive, for your touch

I wish that we could walk together beneath the sun
Oh, what would Barnabas Collins have done

Sometimes...
I've longed for you to long for me
I've longed to get you hard and wet
Should I play it cool for all eternity
Or chase you like the cliff-fallen Josette?

Should I seek our your replacement?
Someone younger, but so much like you?
Someone who would know my excitement
As I force them to do what I wanted you to

Or should I let the pain of life burn me to death?
Take the stake, so sharp and deadly, of bad luck
I really should forget your cold heart and warm breath
Because honestly, my darling, you simply

SUCK!!!

Hungry for you in the darkness, I just want to run
Oh, what would Barnabas Collins have done

SMALL SCREEN CUCKOLD

I wanted to nibble your neck
You wanted to watch *Star Trek*
I longed to kiss your lips
You were obsessed with Susan Powder's hips
I yearned to lick your ears
You preferred to get drunk with *Cheers*

We're a pretty boring team
Neither one of us is a screen dream
I can't walk into a room and make you scream

Not enough leather and lace
Nothing as dramatic and sexy in my face
As in your sleazy Melrose Place

Your TV world is so compelling
Your fantasies I keep smelling
Well, I'm sorry I wasn't created by Aaron Spelling

I don't think we're going to work
You're nothing but a Prime Time Jerk
And I'm never going to be Captain Kirk

MID LIFE CRISIS

You used to ignore me
Now you just bore me
AND IGNORE ME

You used to inflate me
Now you berate me
AND HATE ME

Once my heart
You could catapult
Now my art
You just insult

It doesn't take a wise old shaman
To see we have nothing in common
It doesn't take the crystal ball of a gypsy
To reveal our affair has grown too tipsy

I'm in the mood for a new dance
I'm in need of fresh romance
I'm really hungry for some carousing
And some naughty sexual arousing

By your side, I'm tired and square
As I sit away life on my butt
If only I wasn't so damn scared
If only I had the guts to be a slut

The Ballad(s) Of
SUE & JOE & DANNY & SALLY

Sue kissed her mom and pop good night; she climbed the stairs and doused the light
Her mom and pop would wish her malice if they knew Joe crawls up her trellis
Shortly after she's had her thrill, she wished Mom had let her go on the pill

Joe goes down, feeling really great; he's proving to everyone he's really straight
No one knows about his secret ardor; he wants to dock in Danny's harbor
But Danny says no way, Jose; Danny wants his guy to be all gay

Danny turns Sally's loins to jello; she drops by daily to flirt "Hello!"
She knows how to wink, wiggle and pout; she wants him to go straight and ask her out
Danny ignores her every loaded crack; she just can't wait to scratch his back

Joe works really hard for his money; he buys nice things for his honey
But now he really owes lots of loot because he got hit with a paternity suit
He doesn't know if he'll get a son or daughter; he only knows he's going to do what he oughtta

Sue's folks are pleased she's tying the knot; Joe's are glad his funny ideas he's forgot
Sue's smiling pregnant on her wedding day; Joe's pretty sad about the FamilyValues way
Plenty of rest area husbands for his fascination, but what a mess from one premature ejaculation

A ROUGH BITTER ENDING TO A ONCE BEAUTIFUL LOVE

You left my arms dreaming of glory

You left me here to tell this story

Now you're sitting there, looking so sad

Well...

WIPE YOUR BUTT WITH A STEEL WOOL PAD!!!

THE ODOR OF LOVE

YOU said our love would last forever and ever
That our sweet love would never, ever end
You promised that we would never sever
But I think our sport's car's gone 'round the bend

YOU swore our love would all but touch the sky
That it was really gonna fly and soar
But I think already we've done said good-bye
And clouds above me are gonna really pour

YOU said our love smelled like a sweet, red rose
But all I smell is a stinky fart
I'm all alone with weepy eyes and runny nose
And I've got constipation of the heart

I LEARNED the hard way; I've paid the toll
Don't ever fall in love with an asshole

SCENTS OF STYLE:

THE ODOR OF FASHION

A SIZE EIGHT SONNET

THE hunger for shoes pursues me
I don't know how I can ever stop it
I don't know why the hunger did choose me
To get rid of it, I have to shop it

I'LL grab the car keys, grab the credit card
And fly out the door, high on the flavah
Of shiny leather shoes with heels so hard
Quickly, I rush off to buy out Fayva

I OWN plenty of shoesies and bootsies
And my friends all call me Imelda
When I'm done shopping, sore are my tootsies
And my tired feet are kinda swell-da

HORRIBLE is the net in which I'm caught
I can't wear any of the shoes I've bought

MY VALENTINE'S DAY SWEATER

When Cupid's Day
Rolls 'round each year
I take out
My own wool robe
Of many colors

There's red and blue
And white and gold
All set against
Green background bold
And I tell everyone
All dressed in red
That this is
My Valentine's Day Sweater

The white is for love so pure
That lifts life from mediocrity's sewer

The gold is for the glow and shine
Of laughing lovers' faces, so divine

The red is for the throbbing lust
That rattles bed posts and makes bustiers bust

The blue is for that sweet sad trip
When from sweet love's embrace one must slip

BUT THE GREEN SYMBOLIZES
THE TRUE NATURE OF LOVE
THE JEALOUSY AND ENVY
THE GAMES AND CONTROL
THE POWER TRIPS AND BLEEDING NEED
THE ACHE OF REJECTION'S BITTER SEED
THE HURT AND HATE
THE PAIN AND GRIEF

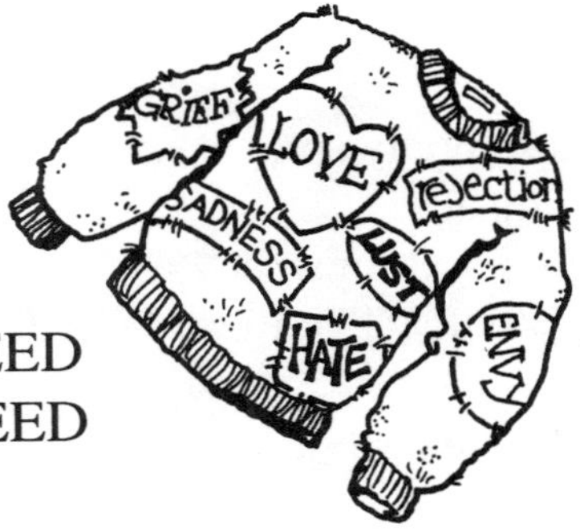

The whole enchilada as rotten and discolored
As a slice of Sunbeam Battered Bread
Left under the kitchen sink
For a month of Sundays

What else should we expect from LOVE?
A little naked boy with arrows
A character who could only appeal
To S&M pedophiles

PRETTY STINKY

Love those pretty spray cans
That promise
Sex appeal and romance
IN PRESSURIZED, FLAMMABLE GAS

It sounds like a squirt of Lysol
Coming out
I smell like
A SUMMER GARDEN GOING OUT

LOVE THAT PRETTY STINKY
LOVE PUTTING ON COLOGNE
'CAUSE IT MAKES ME SMELL
PRETTY STINKY

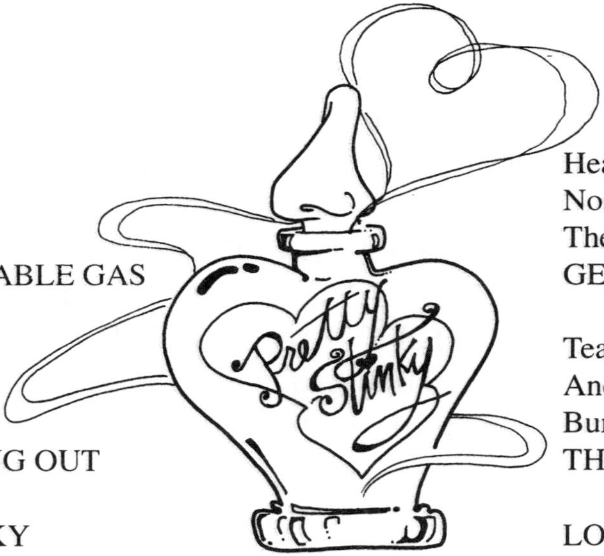

Heads were turning
Noses sniffing
They were wanting to
GET ME DOWN ON THE GROUND

Tear off my clothes
And inhale
Bury their noses in
THOSE PRETTY STINKY SPOTS

LOVE THAT PRETTY STINKY
LOVE PUTTING ON COLOGNE
'CAUSE IT MAKES ME SMELL
PRETTY STINKY

FASHION STATEMENT

Monday morning found it raining. Monday morning found me running. From store to store. Desperately doing errands, while wicked rain leaked through the soles of my shoes. Soaking my socks.

Tuesday morning found it sunny. Tuesday morning found me funny. Falling behind schedule. Fumbling around for clothes, I threw on what I could find. Yesterday's socks, now hard and dry.

Tuesday evening found me blushing. Tuesday evening found everyone rushing. Over an odor powerful enough to kill. Matching their outrage, I pretended not to know the source of the odious perfume that had fouled the room. I kept an eye open for a getaway. So I could scrape off those socks and throw them away.

I KNOW WHO I AM

I slip on my Calvin Klein jeans
Over my Fruit of the Loom panties
I know what Polo or Armani means
I can waltz in and light up the shanties

I love my Red Sox cap
I adore my worn Patriots jersey
I'm great in the Guess that came from the Gap
Smell sweeter in Jovan's Musk than Chocolate from Hershey

With a Marky Mark sneer
A bit of Rolling Stone groovy
I'll grab a can of Adolph Coors beer
And watch a Rob Lowe porno movie

Bring some Jim Carrey faces along
For luck, a Billy Graham prayer
I'll sing a Carly Simon song
And just like Michael Jackson, **"I'LL BE THERE!"**

THE STENCH OF SUCCESS:

$

THE ODOR OF WORK

WORK! WORK! WORK!

WORK! WORK! WORK!
That's all I ever do
Never can get through

Work on my body, work on my life
Work all day, and work all night

Work on my job, work on my career
Work seems to be the only thing here

Work and work, and it never gets done
It may be right, but it ain't fun

Work like this from year to year
and here I am, still stuck here

DAILY PLANNER

Another day, and how I ache
This awful job's too much to take
I've got to quit, for Heaven's sake
Thank God! It's time for coffee break!

Back to my desk, and to the crunch
I hate this job a real bunch
I've got to follow-up on my hunch
Oh, what the hell! It's time for lunch!

Working here, I'm just a gnome
I hate putting up with all their loam
I wish the world I could freely roam
Glory Hallelujah! It's time to go home!

CLOCKWATCH BLUES

Tick-tock, tick-tock
My life in chains; the day padlocked
I sit and work and watch the clock
Tick-tink, tick-tink
The seconds wink; the minutes blink
So slowly by the workdays slink
Tick-teek, tick-teek
It's like this every day each week
From month to month till I could shriek

Everyday I'm in this place, staring at that digital face
A certifiable basket case, I long to win this timeclock race
I work from 9 to 5 till when it's time to race for home---but then
I find myself back here again
As dully predictable as Big Ben

Tick-tock, tick-tock
In my cubicle, my life's in hock

SO MOVE YOUR ASS, YOU STUPID CLOCK!!!

FIVE FIGURE WAGE SLAVE

If the landlord doesn't own me
The mortgage banker will
The credit cards have chained me
Till I'm worn, bent and still
My brand new car is financed
The insurance people peck
The whole damn world's a creditor
That bites me in the neck

This nine-to-five eats up my days
And life really is a BITCH
I spend the best of my best years
Making others RICH
I pick up my pay, I pay my bills
I feel my sanity slip
Whoever thought so many zeroes
Would add up to so much ZIP

T A X I D E R M Y

I work so hard for a petty wage. I'm so full of bitter rage. Your greedy hand is out.
So many forms; such mean and ugly rules. We're all just tariff-ied fools. Thomas Paine, what's it all about?

Another cold letter full of threat. Lord, it always makes me wet my working class bed.
I sign on penalties of perjury. I think I need surgery on my broken, over-taxed head.

What does it all buy, I want to know. Where do my tax dollars go? To those living easy?
Corporate crime arrogantly bleats. Potholes fill our nasty streets. My taxman must become sleazy!

My economic life's a rotten mess
My pride and sanity you undress
More quickly becomes so much less
GOD, I HATE THE IRS!

My spirits you so completely depress
Make a sham of all my success
It's getting so hard to impress
GOD, I HATE THE IRS!

LIKE THE SEVEN DWARFS, ON I PRESS
COME APRIL 15, I REGRESS
I KNOW I'VE LOST THIS GAME OF CHESS
GOD, I HATE THE IRS!

I FEEL POVERTY'S BONY CARESS
I DO NOT WEAR DAME FORTUNE'S DRESS
WHY OH WHY ME--I CAN ONLY GUESS
GOD, I HATE THE IRS!

MR. LOTTERY

Let me be free, Mr. Lottery
Be my lucky ticket
So I can tell them to go stick it

Let me be free, Mr. Lottery
I've spent enough on you
I think some good luck is long past due

Let me be free, Mr. Lottery
I want to give them the axe
And wallow in my winnings
Before the income tax

FREEDOM
That's all I ask
It's out of my grasp
Mr. Lottery, if I tend to beam
It's 'cause you sure are a nice dream

"TONIGHT'S

WINNING

NUMBER

IS..."

Oh well, I'll just keep going on
Bored in hell

FLUORESCENT LIGHTS DON'T TAN

She sits
with a glow on her skin
We all know where she's been
She smiles
With a gleam in her eye
"I wish I could go," we lustfully sigh
She flashes
The tan on her gams
Oh fellow workers, jealousy rams

FOR FIFTY WEEKS
HER BUTT CHEEKS
ARE GLUED TO HER CHAIR
FOR 25 FORTNIGHTS
SHE TYPES, FAXES, FILES & GRIPES
WANTING TO GO HERE...
THERE...
EVERYWHERE...

She returns with a boast
It was all the most
She's brown as wheat toast
We all look like ghosts
Oh, so full of envy
Of this Caribbean dancer
And her sun-baked flirt
With SKIN CANCER

THE "ASS" IN ASSOCIATE

Nightmare Co-Workers

You're in a mood if the sun is shining
You're just as miserable when it rains
You spend your time constantly whining
Moaning and groaning about your pains
Something's always wrong
Something's always rotten
In order to get along
I gotta stuff my ears with cotton

You're here each day, a pathetic mess
Bitching, ranting, storming grit
You're misery's catching, I must confess
Have a stroke when you throw a fit

You're at my desk first thing in the morning
You're on my back last thing at night
You're full of command, demand and warning
And wisdom way beyond my feeble sight
Something's not right
Something's quite rotten
Despite your sting and your bite
Your authority is most misbegotten

We have the same title and pay
and it's just my intolerable luck
To be working in Megalomaniacal Bay
I hope you get hit by a truck

Your out box is always empty
Your in-box always bulges full
You come whining to tempt me
"Help me out!"; such lousy bull
This just isn't right
This is pretty damn rotten
The end of my work's out of sight
Yours you've conveniently forgotten

Your lazy ways ruined so many days
A good mood? You know how to spoil it
Cleaning up after you's put me in a haze
Go drown yourself in the toilet

THE PETER PRINCIPLE

How did you get to where you are?
To explain it, I'm at quite a loss
I could search for answers near and far
But it's all academic because you're my boss

I've grown quite tired of your silly rules, of being just another mule in your pool, a tool on a stool so you can be cruel

I've grown quite weary of your constant lies, your annoyed sighs and phoney alibis, my morale sighs and cries, then dies

I've grown quite old from your incompetence; you sit on a fence, so incredibly dense, hence your office opulence makes no sense

How did you get to where you are?
If I saw the light, I'd definitely repent
The company may think you're a shining star
All I see is a hopeless incompetent

PETER PRINCIPLED OUT

It's said we rise to the level of our incompetence
And I must believe that is surely true
But I swear insanity will be my defense
If I ever have another boss as

ROTTEN

As you

AN OPEN LETTER TO:

Agnetha
Bjorn
Benny
Anni-Frid

I love the song where you lament
How hard it is to pay the rent
But you're rich
Lots of Kroner make an Eden
Even if frigid Sweden
Is Socialist
Here I am, a working man
Going nowhere, kicked in the can
I just can't seem to get ahead
Sometimes I wish that I were dead
ABBA, ABBA, ABBA
I want to grabba
The success you've got
ABBA, ABBA, ABBA
I'm a crabba
'Cause you're what I'm not
Working life is a drag
If I was in **ABBA**
My heart wouldn't sag

Your CDs sell just like they're new
But you haven't played since '82
Such a deal
My paychecks just don't cover it
So sick and tired of all this shit
I could squeal
So stick to Fernando and Napoleon scenes
Super Troupers and Dancing Queens
You don't know what it's like this way
So what can you possibly say
ABBA, ABBA, ABBA
It's your Cadabra
That I want too
ABBA, ABBA, ABBA
I'll catch a cabba
And I'll fly to you
I'll be your fifth member
Won't you let me be in **ABBA**
S'il vous plait, Voulez-Vous?

SITTING IN YOUR OWN PEW:

THE ODOR OF RELIGION

CHILDHOOD MEMORY
FROM A PROTESTANT PAST

I grabbed Grandma by the arm
Pointing, frightened
"Look!" I screamed
I was six that day

There were three of them
Black-robed, old hags
"Witches!" I screamed
Nuns in a shoe store

They looked down at me
Smiling, cooing
"Ooooooh!" I screamed
Their feet sure smelled

QUERIES FOR A STARRY NIGHT

When I look up into the night sky and count the thousands of stars, I think of God's wondrous plans
Like how he made us all equal, although some of us are more equal than others
Like POPES and PRIESTS, REVERENDS and RABBIS
The holy folks who have a direct line to the office of the big guy
And I PONDER, WONDER and MUSE...

Does the POPE POOP?
Do NUNS have to PEE?
Does a RABBI RETCH?
Or a MINISTER MASTURBATE?
Does a SOOTHSAYER SNEEZE
Does a PREACHER wipe his BUTT?

I can't picture a PRIEST PUKING
Or a CARDINAL cutting FARTS
Does a REVEREND pick his NOSE?
Does a BISHOP pick his TEETH?
Does a PASTOR strain on the TOILET
While his WIFE dreams of ORGASMS from the PAPER BOY?

It seems to me God would have spared his special people--the ones who have the right to tell us how to live
--from the burden of daily life and the rigmarole of Q-tips, toothpicks and toilet paper
Oh, I know there are sex scandals, but they use someone else's genitals

FAMILY CHANNEL VALUES

(With no Apologies to Pat Robertson or his little ex-reed Ralph)

There's lots of Disney movies
And wholesome Western fun
Many manly bullets flying--but
You'll see no spilled blood run

You can watch the Three Stooges
Poke someone's eyes out
And Jan and Marcia Brady
Give whitebread life a pout

Enjoy family game shows
And TV has-been jokes
700 Club Extravaganzas
Homeshopping for good, clean, decent folks

There's lots of country music
And Christian singers too
Hell and brimstone preachers
Drive the devil out of you

So when you're really tired
Of sickness, filth and smut
Of ugly, bloody violence
And dirty boobs and butts

Just channel surf to Heaven
Recline and veg your bod
Enjoy cathode ray paradise
Brought to you by a Man of God

"IT'S A GIRL!"

A Major World Religion in an Alternate Universe

They wouldn't let them in the inn
Poor pregnant women just can't win

She had a husband who was no stranger
He made her comfortable in a manger

She screamed in labor; a shocking sight
As a brilliant star lit up the night

Wearing that big yellow ring 'round her head
She looked up from her maternity bed
Smiled and said

"IT'S A GIRL!"

The Angels greeted the Shepherds that day
To tell of the miracle child on the way
The wise men travelled from afar
To bestow the cherub with gifts from a star
But old jealous Herod shook his curls
And swore to kill all the baby girls
She had friends who couldn't be sweeter
They protected the sweet baby Jesusita

And how she shocked the staunch idealogue
When young Jesusita preached in the synagogue
Oh, how divine our Heavenly daughter
As Jesusita danced and pranced across water
She certainly knew how to party divine
Dear Jesusita turned water into wine
With love and joy, she granted wishes
Making a feast from loaves and fishes

And, dear readers, this is as good as it gets
Jesusita recruited her twelve disciplettes
Never was there so fine a scene
As when Jesusita comforted Mario Magdalene
All in Jerusalem was so very calm
When her followers greeted her with leaves of palm
But, oh, the world turned sad and moody
When dear Jesusita was betrayed by Judy

All the world mourned its loss
As sweet Jesusita hung from her cross
But all would be fine on nigh
Jesusita rose from the dead and into the sky
So the world that followed would change
Everything would be quite rearranged
For men, for women, for children too
How different do you think these people will do?

AND ALL BECAUSE --IT WAS A GIRL--THE SWEET AND INNOCENT LITTLE BABY JESUSITA

SHE GOOPS

She GOOPS
and the world GOOPS with her

She GOOPS
and we all GOOP for God

She GOOPS
And the people GOOP for tomorrow

Because she GOOPS
We know tomorrow will come
And be good

She GOOPS
It is the right thing to do

She GOOPS
She is of moral vision

She GOOPS
And the love flows freely

Because she GOOPS
We know we don't need
To ask questions

The strong GOOP freely
The weak GOOP in secret
Do some GOOPING today--just because

GOOP a friend for Christ
GOOP ten percent of your income
GOOP for your heart's sake
GOOP for the good of the country

Don't ask what GOOPING can do for you
Ask what you can do for GOOPING
You deserve a GOOP today
Get a piece of the GOOP
Save a GOOP today
Just for the GOOP of it
Win one for the GOOPER

Ye, though I walk through the shadow of death
I will fear no evil
The GOOP
And The GOOPED
Comfort me
I am GOOPED
GOOPED
GOOPED!!!
Forever safe

GOOP for all its worth
GOOP for America

THE FUMES OF FATE:

THE ODOR OF LIFE

T R O U B L E

SOMETIMES IT SEEMS TO FOLLOW HIM

EVERYWHERE HE GOES

BUT SOMETIMES IT SEEMS LIKE

HE LEADS IT BY THE NOSE

CHRISTMAS DINNER

DON'T let a full plate faze ya
Grab your fork and just dig in
You're eating for all the starving in Asia
So plow in and eat TO WIN

THERE'S plenty of cheeses and lots of cake
Glasses of pink champagne and dry white wine
Be sure to rake in everything you want to take
Fill your plate, grinning a delirious, "MINE!"

LET your cup runneth over. Your plate,
Why just let it spill all over the floor
Don't worry a bit about gaining weight
Just hold it up screaming ecstatically, "MORE!"

TAKE plenty of Aunt Sophie's greasy gravy
There's plenty of oily stuffing from Cousin Maude
There's Swiss chocolates from Uncle Davy
So eat like a pig via THE GRACE OF GOD

HAVE another leg of lamb for the SAVIOR
Eat another chicken breast for the VIRGIN
In the name of the HOLY GHOST, savor the flavor
When you're done chowing, call a SURGEON

CHRISTMAS DINNER II

SOME people think it's yummy
But it's usually pretty scummy
Just eat and eat , then eat some more
Until you cannot move to cross the floor

SOME people think it's hell
But at least it doesn't smell
Just sit there and turn green shades of gray
And hope and pray the feeling goes away

SOME people think it's funny
And it's usually gross and runny
No sweat, no strain
Just sit there and let it drain

SOME people think it's painful
And it's usually pretty rainful
Just groan and spray
All that nasty food away

YES, some people think it's yummy
But it's usually pretty scummy
Just eat and eat until it's gone
And wish to hell you hadn't later on

SLAVES IN TOYLAND

"Our children are the future," every political pundit will say
But we're concerned about the PRESENT--That is, Christmas, Easter and Happy Birthday
Come right in and walk the aisles and let your eyes grow wide and bright
Toys and toys for miles and miles--a child wandering Toyland is a beautiful sight

"LOOK IT!"
Bubbles, jump ropes and superballs
Blue bikes, red wagons and golden trikes
Crayons, frisbees, balloons and baby dolls
Badmitton nets and swingsets for little tikes

"I WANNNNN IT!"
The Power Rangers are mighty slick
Star Wars games, helmets, ships--all on deck
Plenty of toys from the latest Disney flick
Or if you'd prefer, you can go Star Trek

"GIMME!"
Here's Barbie with more faces than Madonna
Get her a car, a house, a matching Ken
Lots of clothes and shoes for her to tryonna
Accessories to satisfy her (your) every yen

"MOMMEEEEEEEEE!"
Get a toy weapon set, you little macho man
Toy dishes, food and make-up for the woman-girl
Etch-a-Sketch will get your junior Picasso in the can
Nintendo will be an action-adventure pearl

"PLEASEEEEEEEEEE!"
Pick some affection from the Cabbage Patch
Cherish ticklish little Elmo and Barney, your purple friend
Mr. Potato Head or Pooh Bear make for a good catch
Take home these buddies; the love will never end

"I'LL HOLD MY BREATH!"
As your love grows, so does our profit margin
So fill your arms and get right there in line
Have Mom or Dad write out a check; we all win
You'll hop, skip and jump back home feeling fine

"I'LL RUN AWAY!"
Our harried clerks and stressed out managers show
Selling toys isn't always very much fun
And think what the tiny toy-making hands know
Playing grown-up in China and Taiwan

"WHAAAAAAAAAAAAAAAAA!"
But if your folks are the happy poor
Well, we don't want to be too graphic
If they haven't got money for you to have more
You little snot-nosed kid, go play in the traffic

CANDYLAND OF OPPORTUNITY

Ghosties and goblins and green-faced old witches
Zombies and vampires; malevolent riches
Make-up and costumes--all Halloween fun
Going on sale under bright dog day sun

Snowflakes and tinsel and big smiling Santies
Green trees and white trees and Jingle Bell panties
More treats and candy than you've ever seen
All goes on sale just before Halloween

AH, THE EARTH SPINS, AND THE YEAR TURNS; HOLIDAY SEASONS NEVER LAST
IF THEY WANT TO MAKE THEIR PROFITS GROW LARGE, THEY'VE GOT TO MOVE REAL FAST

Big hearts of candy and cards promise love
Blessing and joy sent from Cupid above
Cuddly brown teddies to send lover's way
All of it selling before Christmas day

Big chocolate eggies and cute yellow chickies
Big smiling bunnies bring lots of spring trickies
Fake grass, plastic eggs, pastel candy corn
All on the shelves before Valentine's morn

IF YOUR SWEET TOOTH ROTS YOUR MOLARS FROM CHEWING CANDY ALL DAY
YOU CAN BET YOUR ROOT CANAL THAT IT WON'T BE THIS SEASON'S HOLIDAY

SYMPATHY CARDS WE'D REALLY LIKE TO SEND

At this time
Of great personal loss
Words can do so little
To ease the pain
But I hope
With all my heart
Comfort can be found
In this sincere, heartfelt thought

JUST THINK HOW MUCH

MONEY

YOU'RE GOING TO INHERIT!!!

SYMPATHY CARDS WE'D REALLY LIKE TO SEND II

The lifetime of knowing
And loving
A dear member of your family
Is the source
Of so many cherished memories
They can now offer you
Great comfort in this
Your great hour of grief

But loss this large
Can linger for so long
Our thoughts are with you
As you face the future
And we gladly offer up
One humble comforting thought

YOU'LL
NEVER
HAVE TO
CHANGE
GRANDMA'S
CATHETER
AGAIN!!!

SYMPATHY CARDS WE'D REALLY LIKE TO SEND III

At a time like this
We try to think
Of the right words to say
To offer comfort
To give peace
To those who are suffering
Such a grievous loss

As we remember
all the things
We could have said
Or should have said
And wish we could have
One more chance
To show how dearly
Our lost one is loved
May you find comfort in this single loving thought

JUST THINK HOW MUCH

MONEY

YOU'RE SPENDING ON THIS FUNERAL!!!

SYMPATHY CARDS WE'D REALLY LIKE TO SEND IV

We know
Losing a loved one
Has left an emptiness
In your life
A void that only
Time
Fond memories
And prayer
Can fill

As you say good-bye
To one so close
To you
Our thoughts are with you
And we want you to remember
One tender thing

WE HATED HIS GUTS!

WE'RE GLAD HE'S DEAD!!!

SYMPATHY CARDS WE'D REALLY LIKE TO SEND V

Time and tide
Turn and roll
And wait for no one

Sooner or later
We all feel the loss
That only life
and love
Can bring

As you now say good-bye
To the one you have loved
For so long
In so many ways
There's one thought
We want you to hold
Dear to your heart

IF YOU HADN'T BEEN SUCH AN

A S S H O L E,

SHE'D STILL BE ALIVE!!!

WHEN THE MASTER'S AWAY

He picks himself up
and crosses the living room
It's time to to go for a walk
but the Master is not home

He sniffs
He scratches
He fidgets
He must wait
HE MUST

Eyes are on him
He looks up
They're watching him
THE FELINES

They scratch up the rugs
They break the lamps
They claw the chairs
They're daring him
To do what they get away with

He fidgets again
He knows what they expect
He shakes his head
Shame, Guilt and Fear

The Master will not be pleased
But they will
He has to live with them.
The Master is only food and wrath

He scoots over
Lifts his leg
Let's go his spray
And watches it run down the wall

IT'S POLITICALLY CORRECT

ELECTION '96 POST MORTEM

Election gymnastics are swill
I can't stand President Bill
Totally grossed out by Dole
Wouldn't hang earrings on Perot
So much noise and expense, all for nil

The senators are self-serving slugs
The reps are as rancid as bugs
Political diatribe flies sour
More foul by the hour
I hate those smiling "ELECT ME!" thugs

Another scandal exposed--what a BORE
It's all but expected anymore
The Liberty Bell's gone to rust
Give me someone to trust
Give me somebody to vote FOR

ANGUISH ABSTRACTED

(The kind of poem some asshole teacher made you analyze in high school when all you really wanted to do was GET LAID)

There's evil on the sidewalk, and tears on the moon
Sorrow on the doorstep, and silver on the spoon

Love in the afternoon means loneliness at night
Metaphor in cartoon lets misery set you right

Day slips on you like glove, but someone has stolen the night
There are those who fight to love, and those who love to fight

So I've weaved my words of anguish and heartache in rhyme
An old dead love can really smell, and once again it's changing time

THE STENCH FOUND IN THE END

Smell the despair
As it runs in rivers
From my eyes
Flaming tears washing
Away the putrid pain
That coagulates
Into dingleberry lumps
On my shoes

Smell the pain
As it runs in crystal streams
From my nose
Mucus of Misery
That makes everyone
In this fancy restaurant--
Site of my last meal--
Ask for their money back

Smell the anguish
As it drips in steady, crimson drops
From my wrists
Indeed, a bloody, ugly mess
Wallow in the knowledge
This is on your hands
And you'll have to pay
To have this carpet cleaned

"Oh, woe is me; I don't know what I'm gonna do"

NOTHING LEFT TO LIVE FOR

There's filth in the sea
 And poison in the sky
Only trouble is free
The rainforests are all gonna die
I have a beat up old car
I don't have any money
I can't find my shooting star
Only weirdos and creeps
Want to call me honey

There's crime in the streets
And drugs at every door
I've just got to wash my sheets
The Middle East is going to war
The President's still lying
About a new, twisted caper
I feel just like crying
Stop & Shop is out of toilet paper

I used to live Ira Gershwin songs
Now I just sing Christine McVie
Either way they both came out wrong
I could never find the harmony
So I'll just sing "Good-bye to Love"
And stop messing around with my head
There's Plutonium in the space shuttle above
And Karen Carpenter is still dead

BENEDICTION

I HOPE I've touched in you, gentle readers
A bell that chimes with empathy
When it comes to life, we are all bleeders
Let your heart cast shadows of sympathy

WELL, I've let out all my pain through my words
Now, hopefully, I will feel better
The agony flew away like wild birds
As I pounded away at my typewretter

FINALLY finished, I lean back sighing
I wrote out all my pain; that's its finish
When I read my words, I feel like crying
My words are harder to swallow than spinach

I WROTE my poems to keep tears off the floor
When I read them, I feel worse than before

Did this book make you LAUGH?
Did it make you CRY?
Did it make you FEEL?
Did it make you THINK?
OR DID IT JUST MAKE YOU SICK?

Whatever happened, let us hear from you!

Comments can be sent to: T & T Publications, Box 23
Boston, Massachusetts 02124
Or FAX to: (617) 282-5749

ODOR FORM

ADDITIONAL COPIES OF *THE ODOR OF LOVE AND OTHER AROMAS* CAN BE ORDERED DIRECTLY FROM THE PUBLISHER

YES, I want to give the smell that lasts a lifetime to friends, relatives, co-workers, and all the other people I want to get even with!

NAME ____

ADDRESS ____

CITY ____________________ STATE ______ ZIP ________________

DAYTIME TELEPHONE: (____) ____________________________

SEND ________ copies of *THE ODOR OF LOVE AND OTHER AROMAS* at $8.00 per copy.
($7.00 per copy for two to four books) ($5.00 per copy for five or more books)
Send a check or money order to: T & T Publications, Box 23, Boston, Massachusetts 02124

Massachusetts residents, please include 5% sales tax

MICHAEL DUBSON SAGE has had some really dumb jobs in some very high falutin, hoity-toity laces. He has also tried his hand at being a college English professor, but he has never been able to get rough a grammar lesson without snickering about the dangling participle. Born on the cusp of Libra and corpio, he is in a constant state of torment because the Scorpio in him is always seeking vengence, but the ibra can't make up its mind about who to get. He believes there are only two kinds of music in the world: her and Non-Cher. He hopes one day to be able to find everything he's mislaid somewhere in his house.

PAUL THEODORE is the creator of the short lived (three years, not all that short) Comic Strip "Top N' Bottom," which appeared regularly in the publication *Leather Journal.* He is an alumni of the New ngland School of Art and Design and is currently employed (Thank God!) as Creative Director of a small rect marketing firm. His birth sign is Cancer, his favorite color black and his hobbies include needlepoint, coupage and world domination. He is especially fond of cuddly bears, black leather and international ffees.

Back Cover Photography by Robert D. Davis, Boston, Massachusetts